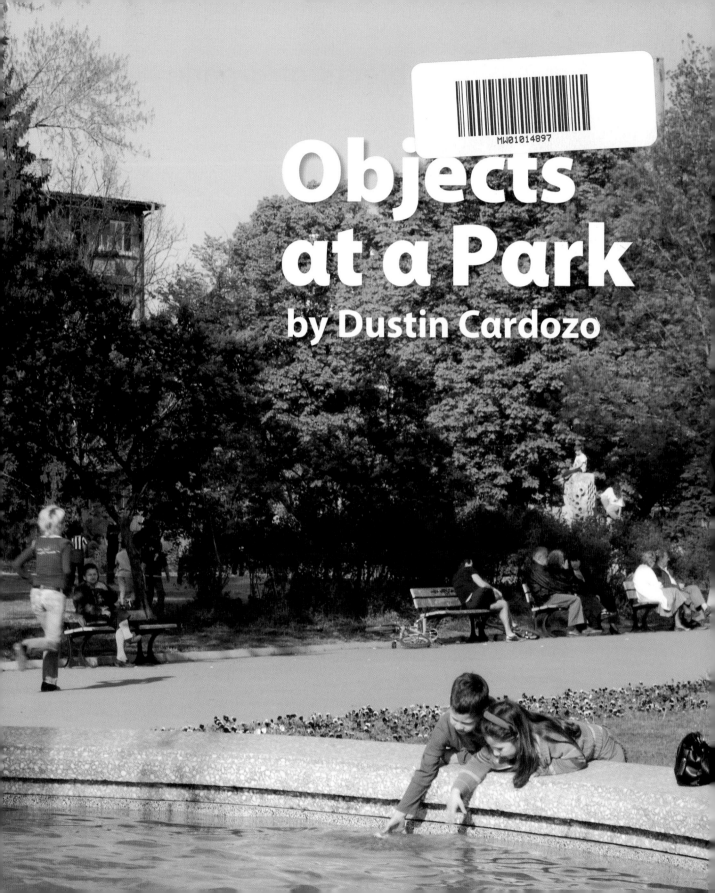

Objects at a Park

by Dustin Cardozo

You can observe **objects** at a park.

You can observe the shape of objects.

A hoop is **round.**

A tic-tac-toe game is **square**.

You can observe the feel of objects.

A basketball is **rough.**

A slide is **smooth**.

You can observe the size of objects.

A bicycle is **big**.

A tricycle is **small**.

You can observe the color of objects.

Some objects are **red.**

Some objects are **green.**

Objects at a Park

round

square

rough

smooth

big

small

red

green